C000216207

Hydraulics in the West

David Cable

Front cover: The bridge near the top of Hemerdon Bank frames No D1024 *Western Huntsman* which needs all its traction power to reach the top with its Penzance–Paddington express. It is August 1964 and the sun shines as it should in summertime. *David Cable*

Back cover: No D7028 is seen from the A38 bridge, leaving Laira shed past the old signalbox. Note the diesel locomotives on what was still the old steam shed, although by this time, April 1964, steam had virtually disappeared from the West. *David Cable*

Previous page: Canton-based 'Hymeks' worked regularly on through trains from Cardiff to Plymouth in the 1960s. In this example, No D7095 is approaching the loops at Hemerdon on a sunny day in February 1964. The 'Hymeks' were a very neat design, embellished by a well-thought-out colour scheme. *David Cable*

Far right: Blue and grey is the order of the day in this shot of an unidentified 'Western' heading the 1255 Newquay to Paddington on 5 July 1969. The train is just about to go under Treffry aqueduct, near the top of the descent from Luxulyan to St Blazey. *Mike Roach*

First published 2011

ISBN 978 0 7110 3540 9

Published by Ian Allan Publishing

an imprint of Ian Allan Publishing Ltd, Hersham, Surrey, KT12 4RG
Printed in England by Ian Allan Printing Ltd, Hersham, Surrey, KT12 4RG

Code: 1105/B1

Distributed in the United States of America and Canada by BookMasters Distribution Services

Visit the Ian Allan Publishing website at www.ianallanpublishing.com

INTRODUCTION

Many words and numerous books have been written about the Western Region's diesel hydraulic locomotives, but this book has set out to try to illustrate the many varied tasks undertaken by the fleet during the period when the two 'Warship' and D6300 classes, as well as the 'Westerns' and 'Hymeks', were in operation. This period lasted from their introduction in 1958 until the withdrawal of the 'Warships' in 1972, by which time the D6300s (Type 2/Class 22) had also disappeared.

Following a couple of introductory shots, the book initially sets the scene in the area predominantly west of Taunton along the main line to Penzance, followed by the plethora of GW branch lines. We then take the Southern lines from east of Exeter to North Devon, Plymouth and North Cornwall. The setting of these interesting classes in such enjoyable surroundings makes for some evocative pictures.

The diesel hydraulic classes operating in this area were primarily the original Type 4 D600 'Warships', unofficially Class 41; the Type 4 D800 'Warships', Classes 42 and 43; the D6300 Class 22; the Type 4 D1000 'Westerns' Class 52 for part of the period; and to a smaller degree the Type 3 D7000 'Hymeks'. The only other hydraulic class, the D9500s, were not known to have ever visited lines beyond Taunton.

The appearance of these locomotives was unique. Each had its own characteristics, some of which appealed to the individual, and some did not. Personally I never liked the look of the D800s (or for that matter the German V200s on which these 'Warships' were based). On the other hand I always enjoyed the North British-designed Class 22s and 41s and the Beyer Peacock 'Hymeks'. The bodywork of the 'Westerns' was fine, but to me was rather spoilt by the strange design of the bogies.

The diesel hydraulics were, of course, initiated by the BR Western Region, following in the footsteps of the Great Western Railway to which all (or at least many) standard features on other railways were anathema. In principle, the use of hydraulic rather than electric transmission gave benefits of weight saving and potential increase in load haulage. The original locomotives did give some problems in their early days, as did several other designs on the rest of the BR system, but were nevertheless assigned to some top-level duties. However, in the same way that these faults were overcome elsewhere by increased knowledge of operating and maintaining the machines, or in some cases fitting different diesel engines, the same results would probably have been achieved with the hydraulic classes over a period had not the BR hierarchy taken anti-WR decisions to standardise on electric transmission.

Technical details

It is not the intention for this book to describe the locomotives in detail — many publications and articles have covered this in depth but a simple summary may assist the reader:

Class	Numbers	HP	Wheel Arrangement	Weight	Diesel Engines
22	D6300-05	1000	B-B	68T	MAN
22	D6306-57	1100	B-B	65T	MAN
35	D7000-7100	1740	B-B	74T	Maybach
41	D600-04	2x1000	A1A-A1A	117T	MAN
42	D800-02	2x1056	B-B	79.5T	Maybach
42	D803-32	2x1135	B-B	79.5T	Maybach
42	D866-71	2x1135	B-B	79.5T	Maybach
43	D833-65	2x1135	B-B	79.5T	MAN
52	D1000-73	2x1380	C-C	108T	Maybach

Class 42 D830 was fitted with Paxman Ventura engines instead of Maybach for experimental purposes.

The Class 22s, Class 41s and the D833-865 series of 'Warships' were built by the North British Locomotive Company, the Class 35 'Hymeks' by Beyer Peacock, and the remaining classes by British Rail workshops at Swindon and Crewe.

The 'Warship' and 'Western' classes were normally scheduled for 90mph operation, but various engines had been recorded reaching 100mph on occasions.

Locomotive operation

Perhaps it was naïve of the Western Region management to order untested diesel hydraulic locomotives when the other Modernisation Plan engines were all diesel electrics. It was also perhaps unfortunate that the North British Locomotive Company was the manufacturer for the initial batches. But a balanced assessment of locomotives which have had a bad, but perhaps unjustified, reputation should take account of factors which affected their day-to-day operations; and let us not forget that they ran the train services throughout the West pretty reliably for several years, so they cannot have been that bad.

The factors include the design of the whole locomotive, including the diesel engine, the transmission, train heating boilers (which themselves had an unenviable reputation), and the control gear, plus manufacturing problems allied to a workforce used to building steam engines with less critical tolerances, using obsolescent machinery and tooling from pre-World War 2 with consequent relatively primitive fitting and assembly techniques. (Remember that the UK was very late in updating its manufacturing facilities after the war, whereas Germany benefited from the postwar situation and Marshall Plan aid to be able to build the V200s with modern facilities.)

Totally new knowledge of the machines and their operation had to be inculcated into an often elderly staff of footplatemen and running shed personnel, some of whom found difficulty in absorbing this new knowledge, working initially in quite unsuitable steam locomotive depots, with all the filth involved, until the new diesel depots became available, and perhaps with inadequate training in identifying faults, especially before a problem caused a failure.

Maintenance of the locomotives allocated to the West Country was mainly undertaken at Laira steam depot in the early days, with limited servicing at St Blazey, Long Rock and Newton Abbot. In 1960, building of the new diesel depot at Laira commenced, and some maintenance took place in the part-finished building until the full facility was in operation in March 1962. St Blazey had its own allocation of locomotives from 1960, and Newton Abbot shed was rebuilt in 1962, so as to take on responsibility for daily maintenance of the 'Warships' used on the Exeter to Waterloo services. Only after these modern facilities were fully activated was it possible for what could be temperamental units to be properly looked after in appropriate conditions, by which time their reputation had suffered. Had all these factors been dealt with as happened with the majority of the less satisfactory diesel electric designs, I have no doubt that reliable units would have been in operation for much longer. But, of course, this was flying in the face of the edicts on standardisation from on high.

Not every line in the West saw diesel hydraulic locomotives at work, even on an irregular basis. With the cessation of steam, passenger services were often replaced by diesel multiple-units, and virtually all wagon load freight had disappeared from most of the branches. The Exe Valley and Teign Valley branches, the lines to Moretonhampstead, Ashburton, Brixham, Turnchapel, Yealmpton and Callington, the branch from Coombe Junction to Looe, Tavistock Junction to Launceston plus the Princetown branch, and the North Cornwall line from Halwill Junction to Wadebridge do not seem to have had any regular, if any, hydraulic operations apart possibly from when track was lifted. The Bude branch through Halwill was operated by a Class 22 for only one year, and apart from occasional runs and a spell of Saturday-only through carriage workings to and from Paddington, the Chacewater to Newquay line was primarily a DMU preserve. The Wenford branch was banned to diesel locomotives larger than a Class 08.

On all the other main, secondary main and branch lines west of Taunton and Seaton Junction the D6300s were seen, and many also hosted Bo-Bo 'Warships'. The original 'Warships' of Class 41 worked the GW main line, but were relegated to china clay workings in Cornwall towards the end. The 'Westerns' worked the main and secondary main lines, but not the Southern main line east from Exeter, and were also able to work lines such as from Lostwithiel to Fowey. The least common class was the 'Hymeks', but they had regular work east from Plymouth to and from South Wales, covering the Southern line from Exeter to Plymouth, the Kingswear branch, and especially the Taunton and Minehead to Barnstaple/Ilfracombe lines. They had also been seen on the Hemyock branch and on the SR South Devon branches.

What has been surprising in looking at the photographs is the number of trains with incorrect headcodes, particularly those hauled by the Class 22s. I have been told that changing the codes was sufficiently difficult for the drivers to manipulate that they didn't bother, leaving the previous code(s) in situ. What train-regulating signalmen thought of this is not known, but passengers (unlike Southern Region commuters) probably had no idea of any codes, let alone the correct ones! However, as far as possible, I have tried to identify the correct working from the appropriate working time tables.

Photographs and acknowledgements

The period covered by this book, 1958-1972, was one during the earlier years of which a good amount of steam working still prevailed, and those photographers who were resident in or visited the West of England often concentrated on these trains instead of the unloved diesels. Well, in view of their relative rarity, perhaps they were mistaken.

However, there were a few enlightened souls, of whom I am pleased to have been one, who managed to take pictures of the hydraulics during this era, and I have been gratified by their willingness to give me access to their valuable collections. I have been surprised by some of the locations they visited, and equally surprised that some lines seem to have been largely ignored. As far as possible, previously unpublished photographs have been used, both in black and white and in colour. The reader will realise, however, that not in all cases is the quality of those pictures up to today's high standards, being taken in a period 50 years ago when disposable income, equipment and film available to the average person were not at the same level as today. But because they are all now irreplaceable, quality has sometimes taken a back seat so as to include a picture in a location which now no longer exists, or a working which reflects that era.

Photographers are credited as appropriate, and it is due only to lack of space that all their submissions cannot be included. To this end, I have concentrated as far as possible on pictures taken in the less usual locations and particularly on lines now closed or completely rebuilt. So pictures at Dainton and between Starcross and Newton Abbot only make a limited appearance, since this stretch of line is, shall we say, rather hackneyed (sic)! With a couple of exceptions, they are all taken in the 15 years the book encompasses.

To these photographers, my appreciation for their help in giving me the opportunity to remind myself of the few years when I lived in Plymouth and Tavistock and like so many, never took as many photos as I should have done. I am sure that the readers of this book will also appreciate their contribution to history. Thank you, gentlemen.

I would like to pay special tribute to Bryan Gibson of Plymouth. Not only did he contribute some very enjoyable pictures, especially of the Kingsbridge branch, he also researched many of his copies of working time tables from the period to identify particular workings from the headcodes, even if only to prove they were wrong!

As ever, the Ian Allan Publishing staff have given welcome support and their help is always appreciated.

David Cable
Hartley Wintney
2010

Great Western Main Line

On the main line from Paddington to the West Country is Langport, near which is the village of Curry Rivel, noted for the sale of wicker products. Heading west in August 1959 is 'Warship' No D804 *Avenger* with an express for Penzance, carrying the headcode for the 1530 from Paddington. However, the shot was taken in the early afternoon, so probably should have read 430 for the 1130 down. From the other direction, No D601 *Ark Royal* approaches with the up 'Cornish Riviera'.
David Cable

Left: 1C41, the 1406 Cardiff to Plymouth, enters Taunton on the down main sometime in 1963, behind an unidentified 'Warship'. Since the distant is not off, the train will be stopping. The full selection of semaphores on view lasted for many years ahead. *Classic Traction*

Below: A Paignton to Bradford express, 1E37, enters the station at Taunton on 27 April 1970. 'Warship' No 811 *Daring* leads 'Peak' No 78, having passed under the superb signal gantry guarding the west end of the station complex. Since these two locomotives could not work in multiple, it is likely that the 'Peak' had failed. Various vans occupy the sidings on the left. As was the wont of the signalmen here, the signals have been returned to danger whilst the train is still under the gantry. *Bernard Mills*

Right: Taking the down goods loop at Norton Fitzwarren with an unidentified freight, 6C60, No D1050 *Western Ruler* is seen in May 1965. The full range of lower quadrant semaphores protects the up main line and Minehead branch, and an up express is due. *Colour-Rail (207180)*

6C60

‘Hymek’ No D7013 leaves Whiteball Tunnel and sweeps down the bank in the footsteps of *City of Truro* towards Wellington with unidentified 1E67. The date is June 1962. *Colour-Rail (31426)*

Left: No D834 *Pathfinder* starts the run down from Whiteball tunnel to Burlescombe and Tiverton Junction with an old-fashioned freight train. The maroon livery, which was carried by quite a number of 'Warships', has not lasted well through the washing plants, particularly that at Laira, which had a reputation that can only be described as dreadful. The date is 29 November 1969, and 8C28 is the 0655 Severn Tunnel Junction to Exeter Riverside. *Mike Roach*

Left: Maroon 'Warship' No D839 *Relentless* looks pretty clean — the Laira washing plant hasn't got to it yet! — as it heads along the down main line in June 1967 at Tiverton Junction with 6C13, the 1550 Bristol West Depot to Exeter Riverside. The station nameboard has lost its signs for passenger connections to Tiverton and Hemyock. *Colour-Rail (DE1835)*

Spending a day at Tiverton Junction in August 1960, taking photos of steam workings, I am very glad to have taken this shot of original 'Warship' No D602 *Bulldog* speeding through with a Penzance to Manchester express complete with TPO coach behind the engine. Although these locomotives had a poor reputation in later years, the fact that they were entrusted to workings such as this indicates that they can't have been as bad as all that. *David Cable*

Left: No D6333 is working the Hemyock and Tiverton Junction pick up goods on 2 May 1968, returning to Exeter wrong line between Stoke Canon and Cowley Bridge. It is running over a temporary crossover; note the fixed home signal. The headcode is wrong; 2C90 is the code for a local train from Newton Abbot to Paignton. Drivers had great difficulty changing the headcodes due to design restrictions on this class, and it was easier to leave an old one in situ. *Maurice Dart*

Left: At the classic location of Cowley Bridge junction, 'Hymek' No D7016 passes with a cement train of Presflos from the Blue Circle works at Westbury to their depot in the yard at Exeter Central. It is May 1965 when all was well, but now not only have the 'Hymeks' and Presflos gone, so have the Blue Circle works and depot. Such is progress! *David Cable*

Left: No D6301 accelerates away from Exeter St Davids to climb the bank up to Central with a train from Plymouth. The photo is undated, but judging from the condition of the engine cannot have been too long after the introduction into service of the engine in 1958. The spotters on the platform must now all be in their 60s or 70s! *Maurice Dart*

Above: An original 'Warship' on the Western Region's prestige train, the 'Cornish Riviera', seen here between Powderham and Starcross in August 1958. No D601 *Ark Royal* is in charge, and shortly afterwards, No D600 *Active* passed with the up equivalent working. The shot is not helped by very grainy Tri X film, but is of sufficient interest to include it. Again, the picture stresses the use of an engine with a poor reputation on a top-class working, and the fact that they were assigned to such duties tends to belie this reputation. Perhaps it was the maintenance in poor working conditions, rather than the design, which was the cause. *David Cable*

Left: No D6333 stands at the exit of the down line at Dawlish Warren on 7 October 1959 with the crew of the engine giving the eye to the photographer, and the personnel in the inspection saloon giving their attention elsewhere. The Laira washing plant doesn't seem to have helped the look of this nice little engine. *Richard Lewis*

Right: Calling in briefly at Dawlish one day in May 1963, I grabbed this shot of 'Hymek' No D7027 hauling a westbound freight along the sea wall. This was a rare working for this class in this part of the world, so I was glad to have had the camera with me. 6C16 has not been identified. *David Cable*

Right: On a beautiful afternoon in 1968/9, No D804 *Avenger* heads away from its stop at Dawlish with an unspecified local working. The tide is well in and notice how few cars are parked compared to the present day. *Bernard Mills*

Left: Heading away from Newton Abbot is 'Hymek' No D7050 piloted by No D6327 with the Kingswear portion of the winter timetabled 'Cornish Riviera' which it will connect to the Penzance portion at Exeter St Davids. The picture was taken in September 1963 when steam was fast disappearing from the West Country. *David Cable*

Left: No D6320 passes the semaphores at Hackney, guarding the approach to Newton Abbot in June 1967, with down freight 9J44, which must be a local trip working or the wrong code. *Colour-Rail (DE614)*

Left: On 28 June 1965, No D6320 leaves Newton Abbot with the Kingswear portion of the 1230 from Paddington. Almost the whole station layout can be seen in the days before the axe fell to reduce such stations to very basic facilities. *Bryan Gibson*

Left: A lengthy up freight enters Newton Abbot from the west behind No D6332 which shows a light engine marker disc! The date is June 1962, and all signals are at attention, having been reset by the signalman in the West box. *Colour-Rail (31791)*

Above: June 1969 sees the 1520 Penzance to Paddington speeding through Aller Junction before the layout was modified to avoid conflicting movements between the down Plymouth and up Torquay lines. The 'Warships' are No D822 *Hercules* and No D869 *Zest*. Note the continental destination boards below the coach windows, a sensible feature for passengers, which lasted for only a short time. *Leslie Riley*

Left: No D845 *Sprightly* heads a train of empty wagons for the quarries at Stoneycombe on 14 July 1969. The refuge provided a welcome place for signalmen at Aller Junction to regulate services up the climb to Dainton from the east. Since Stoneycombe was halfway up the bank, no doubt the driver had been told to get the engine to live up to its name. *G. F. Gillham*

Right: Dusk is falling as No D1065 *Western Consort* approaches Totnes with the inaugural 'Golden Hind', the 1720 from Paddington to Plymouth, on Monday 15 June 1964. Those hoping for a headboard on the front were disappointed. *Bryan Gibson*

Right: At the junction for the Ashburton branch at the east end of Totnes station, No D6339 runs round a weedkilling train, prior to propelling it onto the branch where the Dart Valley Railway would take over. It is not thought that any diesel hydraulics worked the branch itself. The date is 14 May 1966 and the down sidings still earn their keep. *Bryan Gibson*

Above: Not many photos were ever taken at this location, the Totnes Quay branch, but on Saturday 5 February 1964, it was playing host to No D832 *Onslaught* which had a train of timber in its care. Even rarer were workings into the cattle dock sidings at Totnes market. *Bryan Gibson*

Left: A somewhat rare coupling of locomotives in the South West, 'Hymek' No D7044 and a 'Warship' bring a down train past Kingsbridge Junction into Brent in September 1963. An unknown Class 22 waits for business to serve the Kingsbridge branch. The photographer was fortunate that his arrival from Kingsbridge neatly coincided with the main line service, which has not been possible to identify. *Classic Traction*

This was a fortuitous shot, the train appearing just as I was about to drive over the bridge. The location is Tigley, on Rattery bank, the signalbox being visible in the distance. No D6336 is hauling what I recorded as the Newton Abbot breakdown crane, but the destination and reason for the trip are unknown. The date is September 1963. *David Cable*

Left: September 1966 at the eastern end of Ivybridge, where D1006 *Western Stalwart* approaches with an express from Paddington to Plymouth. Of special note is the marine-type circular windscreen wiper fitted to this and D1039, but presumably this was a failure since they were removed quite soon after. *David Cable*

Left: At the same location, but looking the other way, one of the seven green-liveried 'Westerns', No D1037 *Western Empress*, heads towards Paddington with its train from Plymouth. *David Cable*

The loop at the west of Ivybridge sees No D1027 *Western Lancer* enter with an engineers' train of what looks like cable-laying equipment. The date is 9 September 1973. *Classic Traction*

This view of Cornwood viaduct is now hemmed in by vegetation, so it is nice to see it in full view with an all blue and grey consist headed by an unknown 'Western' on an unrecorded date. 1A79 is a service to Paddington. *Classic Traction*

Above: The up main is set for a clear run through for the 1010 St Blazey to Severn Tunnel Junction freight. Both 'Western' and date are unrecorded. *Classic Traction*

Right: When the 'Westerns' were first introduced they were entrusted with top mainline duties, but not initially on their own over the South Devon banks. In June 1963, the 'Cornish Riviera' is about to descend Hemerdon bank in the charge of No D1039 *Western King*, piloted by Class 22 No D6329. *David Cable*

People often say that diesels do not have the same impact that steam has. Well, I can only say that an experience such as this, with the vibration through the feet as a 'Western' comes under the bridge at the top of Hemerdon bank under full throttle, was, to say the least, exhilarating! The train is a through Falmouth to Paddington express made up at Plymouth to 14 coaches, and No D1039 *Western King* has its work cut out to reach the summit. It is August 1964, and a year earlier, a Class 22 would have piloted the 'Western' on such a train. *David Cable*

Right: No D1032 *Western Marksman* is seen on 28 December 1970 at St Mary's Bridge, Plympton, with 1Z03, a Plymouth to Paddington Holiday Inn return Christmas Charter. The operating department would not trust the Blue Pullman to work on its own over the South Devon banks, so locomotive assistance was provided to Newton Abbot. A very interesting shot of a very rare working indeed. *Bernard Mills*

Right: At the foot of Hemerdon bank was Plympton station, the remains of which are seen here on 6 September 1965, as No D6305 trundles past with a well-loaded down freight possibly from Ivybridge. *Bryan Gibson*

Left: In the days before road traffic became excessive, the bridge at Tavistock Junction yard provided a good location for observing train workings. In this view, 'Warship' No D845 *Sprightly* heads west towards Plymouth in June 1963 with a summer Saturday 0755 Carmarthen to Penzance train. Note the unusual treatment of the yellow warning panel, and what appear to be yellow window frames, which have disappeared in the earlier shot (see page 17) of this engine at Aller Junction. *David Cable*

Left: Not many shots are seen of a Class 22 close up from above, so to put one on record here is No D6302 shunting the yard at Tavistock Junction in May 1964. Twin headcode panels have been added by this date to one of the original engines of this class. *David Cable*

Right: With the Tavistock Junction yard full of wagons on both sides of the main line, the up 'Cornish Riviera' has been entrusted to No D866 *Zebra* on this day in May 1964. Of course, subsequent rationalisation has reduced everything to a mere shell of what used to be. *David Cable*

Right: A different view of the yard at Tavistock Junction, seen here through a telescopic lens at the west end. The date is 16 October 1973, and the unidentified 'Western' is hauling 2B68, described as a salt train. An interesting shot from a different viewpoint, which in the current surroundings on the A38 bridge would be somewhat risky! *Classic Traction*

Left: On the approaches to Laira, following the River Plym, No D815 *Druid* is about to pass under the original A38 main road into Plymouth. The train is the down 'Cornish Riviera', the date is April 1964 and semaphores still abound. *David Cable*

Left: No D7028 is seen from the A38 bridge, leaving Laira shed past the old signalbox which amongst other things used to control the right of way of the Lee Moor tramway. It is hard to picture in this day and age the Great Western main line giving way to a horse with wagons of china clay! Note the diesel locomotives on what was still the old steam shed, although by this time, April 1964, steam had virtually disappeared from the West. *David Cable*

Mount Gould Junction was where the Southern line to Plymouth Friary joined the line that ran along the eastern side of Laira shed. In this picture, taken on 13 June 1964, 'Hymek' No D7089 brings the empty stock of the 1516 Okehampton to Plymouth for stabling in the remains of Friary station. The new diesel maintenance depot at Laira is now in operation. *Bryan Gibson*

Left: No D824 *Highflyer* flies into Plymouth North Road in May 1971 with a lengthy freight train comprising mainly vans, no doubt carrying produce from Cornish farms. The high viewpoint gives a good chance to see the layout at the western end of the station, with the, by that time, unused branch to Millbay veering off to the left. *Bernard Mills*

Left: Sometime in 1963, the photographer recorded No D7028 and an unknown 'Warship' departing Millbay. This was certainly a place where not many shots were taken in diesel days. Train 1M91 is the 0730 ECS Millbay to Plymouth North Road, and then the 0900 to Crewe. *Classic Traction*

Above: Action at Devonport Junction on 7 March 1968. 'Warship' No D839 *Relentless* on its way from Penzance with the 0755 to Paddington, sweeps past D1023 *Western Fusilier* which waits for the road after stabling royal working 1Z02 at Devonport Kings Road. *Bernard Mills*

Right: It is nice to see the odd shot of a freight with a 'Western' in a less common location. No D1010 *Western Campaigner* has come off the viaduct in the background and is entering Keyham station near Devonport on 8 May 1973 with 6B59, the 1505 Ponsandane to Exeter Riverside. *Classic Traction*

No D6312 brings an up engineers' special working off the Royal Albert Bridge on 19 April 1962. The train has an interesting selection of vehicles, the third and fourth looking of particular interest to rolling-stock connoisseurs. *Brian Haresnape*

The up 'Cornish Riviera' passes the banner signals at Saltash on Sunday 29 April 1962. 'Warship' No D839 *Relentless* carries both reporting number and headboard. Note the milk tanks in the small goods yard and the various buildings, the stone wall and signals, making an attractive collection of features for the model maker. An interesting item at Saltash was the position of the water crane on the down platform, close to the footbridge in the centre of the station. This was so that pannier tanks in the centre of push-pull services from Plymouth could take water without having to move down to the end, with trailer coaches being off the end of the platform. *Brian Haresnape*

Left: On 3 May 1964, I had gone out to take 'The Cornubian' on its return from Penzance hauled by 'West Country' No 34002 *Salisbury*: Whilst waiting in rather poor conditions, I grabbed this shot of No D801 *Vanguard* passing the old station site at Defiance with the long-lost milk traffic from St Erth to Kensington Olympia. Note that the down loop was still in use at this time. *David Cable*.

Left: St Germans station on 7 June 1969 sees No D834 *Pathfinder*, which we saw at Whiteball bringing seven milk tanks along the up line with 6A14, the St Erth to Kensington Olympia service. The old Great Western station sign still commands attention and gas lamps are still the order of the day. *Mike Roach*

Right: A view at Liskeard where, except for the train and station signage, nothing has changed since this photo was taken in 1962. No D6319 has arrived with a local freight and is about to shunt back onto the line which connects with the branch to Looe. The driver leans out to watch the shunter's hand signals. *Rail Photoprints (AHB810)*

Right: Class 22 No D6323 whiles away the time as it shunts the goods depot at Liskeard in November 1970. Headcode 8C59 is for the 2330 Truro to Tavistock Junction, so either it is wrong or the train is *very* late. A few coal wagons occupy the sidings and since no one is in the cab it must be time for a cuppa. Note the three (not ten) green buses standing on the wall, which presumably all arrived at once! *Bernard Mills*

The east end of Bodmin Road station shows the gantry for carrying the water main to the up platform, but by this date, 2 April 1964, is hardly ever used since steam had virtually disappeared from the West of England. The unidentified 'Warship' heads a train for Paddington, and the rear of a train to Padstow can be seen at the branch line platform. *Kidderminster Rail Museum (019036) / V. R. Webster*

The superb station nameboard enhances this view of No D601 *Ark Royal* entering Bodmin Road on 27 August 1964, with a train from Plymouth to Penzance. Although the headcode shows this to be an express, one disc is so dirty that the train might be mistaken for a local freight!
Kidderminster Rail Museum (090595) / P. J. Lynch

Left: Coming into Par from the St Blazey direction is a freight train carrying predominantly china clay products. It has almost certainly come down from the workings around Bugle and Roche behind the unidentified 'Warship'. The photo was taken some time in 1970. What the signal in the off position is indicating is dubious, since the through line is single at this point. *Kidderminster Rail Museum (041757) / D. Wittamore*

Left: A pair of 63s, Nos D6319 and D6336, approach St Austell with a train of china clay headcoded 6C40, but since this photo was obviously taken in daylight and the code refers to the 2255 Tavistock Junction to Truro freight (due through St Austell non-stop at 0107) and the train is going in the opposite direction, the code must have been left unchanged from a previous occasion. The date is 8 October 1968, and split headcode boxes are now in fashion. *Rail Photoprints (081068)*

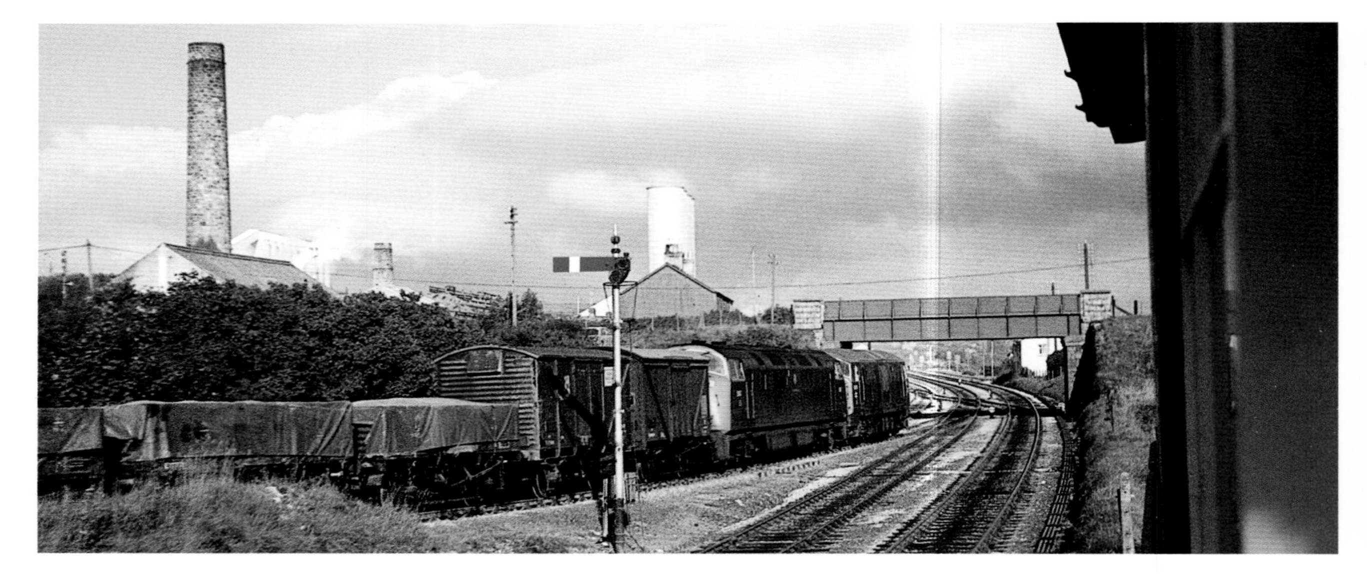

Right: D817 *Foxhound* piloted by Class 22 No D6318 is leaving the branch from St Dennis Junction at Burngullow with a load of clay from either Drinnick Mill or Parkandillack, making its way to Par and probably Fowey. The tower of the Blackpool drying plant and an old chimney dominate the skyline. *George Hammett*

Right: With a mixture of chocolate and cream and maroon stock, train 1C30 heads west behind a 'Warship' on an unknown date. The location is between Coombe St Stephens and the River Fal viaduct, between St Austell and Truro. The impact of the local china clay industry on the landscape is all too obvious in the background. *Eric Irons*

Left: On 17 October 1961, the British India Line chartered a Pullman train to convey passengers from Paddington to Falmouth in conjunction with the refurbishment of SS *Nevasa* as an educational cruise ship. The train is seen entering Truro station. Note the profusion of sidings as compared with today's layout, and the early postwar Farina-bodied Austin A40. *Eric Irons*

Left: Truro station is host to No D604 *Cossack* which is heading an up parcels service on 23 August 1965. The full details of the design and colour scheme for this class of five locomotives are clearly seen in this shot, including the later addition of a half front-end yellow panel. This locomotive survived until September 1968, when it was cut up at Cashmores in Newport. *Richard Lewis*

Right: Penwithers Junction is where the branch line to Falmouth leaves the main line west of Truro. No D1008 *Western Harrier* is seen coming up from Penzance on 20 February 1970 with an express for Paddington. Note how straight the branch line is at this point, plus the signals and signalbox. Some ballasting has taken place on the down line, and an errant drainpipe has been left behind. *Bernard Mills*

Right: A pair of '6300s', namely No D6300 and No D6302, are in charge of one of Western Region's prestige trains, the up 'Royal Duchy', which is seen shortly after passing Chacewater station in May 1959. Two clean engines and a rake of chocolate and cream coaches blend in very nicely with the Cornish countryside. *Michael Mensing*

The little station at Scorrier between Chacewater and Redruth is no more, although the chapel on the left is still in situ. From the footbridge we see No D819 *Goliath* with a westbound parcels train entering the down platform, whilst from the up platform can be seen No D6348 following on with a down local freight. The date is June 1963. *David Cable*

Above: Redruth station is not one of the easiest at which to take pictures at the London end, normally requiring a rather cramped head-on shot as seen here. No D1021 *Western Cavalier* is starting to accelerate away with the up 'Cornish Riviera' which left Penzance at 1015. Steam heating is working and needed since it is Boxing Day 1968. The station forecourt displays an interesting selection of vehicles. *Rail Photoprints (261268)*

Above: Running alongside the wall of the Holman Brothers factory at Camborne, 'Warship' No D836 *Powerful* is about to negotiate the level crossing at the east end of the station as it brings the 0730 Paddington to Penzance to a halt. It is Saturday 24 July 1965, so the train is probably full of holidaymakers. *Bryan Gibson*

Below: No D601 *Ark Royal* enters Camborne on a wet evening in July 1958 with the 1920 Penzance to Plymouth service. *Maurice Dart*

Left: Only two of the 'D800' hydraulics were not named after warships — Nos D800 and D812. The former, named *Sir Brian Robertson* after the BR Chairman at that time, is seen arriving at Gwinear Road with the down 'Cornish Riviera' to make the connection with the Helston branch train. The date is 22 May 1959. The locomotive carries the later style of named headboard designed for use with steam locomotives, as well as the reporting numbers using steam-style metal plates, which were the only type available for the first three of this class when built. Note the destinations on the station nameboard, most of which had to be reached by bus! *Michael Mensing*

Left: The train standing at Hayle Wharf on 14 July 1971 is headed by 'Warship' No 810 *Cockade*. 7B30 is thought to be a trip code for working between Hayle Wharf and Penzance. Note the two men probably going to work in the then traditional manner of cycling and wearing caps. *Exe Rail*

Right: At St Erth, the junction for the St Ives branch, No D6315 clatters over the points on its way into the branch. The track layout here has changed substantially since this picture was taken in June 1963: the sidings on the right have gone and rail access to the milk depot has ceased, but at present semaphore signals remain. *David Cable*

Below: No D601 *Ark Royal* appears several times in this book, and in this case is passing Marazion with the up 'Cornish Riviera', which also features many times. The neat little station seems to have a footbridge cast at the old Southern Railway concrete plant at Exmouth Junction. The well-known camping coaches are not visible in this undated view, which must be in the earlier days of the 'D600s' being used on the premier West Country expresses.
Colour-Rail (204270)

Left: 1M99 is the 1245 Monday-Friday or 1315 Saturday-only Penzance to Crewe Passenger and Mails, with through coaches to Liverpool, which has left behind No D841 *Roebuck* in September 1964. The water tank at Long Rock shed stands proudly at the top of the headshunt on the right. *Colour-Rail (DE1689)*

Left: The tide is out at Penzance, where No D866 *Zebra* runs over the points and into the terminus with 2C20, the 0900 stopping train from Plymouth. It is 1965, and another scene where little has changed apart from the rolling stock. *Rail Photoprints (081965)*

Great Western Branches

Minehead Branch

Right: An unidentified 'Hymek' is seen near Washford, Somerset, with the 1115 Minehead to Paddington through working on 22 June 1963. *W. G. Sumner*

Right: Stogumber on the Minehead branch sees a pair of 'Hymeks' passing a well-positioned caravan for a train enthusiast. Nos 7100 and 7011 are in charge of a train from Minehead to Paddington train, although the 1C17 headcode indicates, the 0630 SO Oxford to Minehead. The date is 28 August 1970. *Bernard Mills*

Taunton–Barnstaple

Whether 3C94 is the correct headcode for an engine and brake van is a matter for conjecture, but No D6337 wears it proudly as it leaves the tunnel and enters Venn Cross on the border between Somerset and Devon on the line from Taunton to Barnstaple. The picture was taken in 1965.
Michael Mensing

Exe Valley Branch

The Exe Valley branch was one of those lines on which steam was replaced by diesel multiple-units until 1963 when the line closed. Freight had ceased before then, so locomotives were virtually unknown on the branches between Exeter St Davids and Morebath Junction Halt, or Tiverton to Tiverton Junction. However, one locomotive-hauled train is seen here as No D6318 brings the stock of the 1520 SO train to Exeter St Davids into Bampton station on 28 September 1963.
Hugh Ballantyne

Hemyock Branch

Right: Whether the patrons of the Culm Valley Inn will turn out to see No D6352 pass with the milk train at Hemyock is uncertain, but the blue-liveried locomotive has 8C07 in tow. The shunter watches over the level crossing in this 1970 view. *Colour-Rail (200928)*

Below: The milk tanks in Hemyock station yard are being marshalled by No D6350 on 15 June 1970. 6C24 is the wrong headcode, being the 1640 Torrington to Exeter St Davids milk and freight. How many lorries are now needed to convey this number of milk tanks? So much for progress and environmental pollution. *John M. Boyes*

Above: No D1006 *Western Venturer* heads towards Newton Abbot with a weedkilling train. The location is Heathfield where the Moretonhampstead and Teign Valley branches met. The date has not been recorded, but apart from a small bush by the platform edge the station looks reasonably well kept. *Bernard Mills*

Right: Blue-liveried Class 22 No 6308 approaches the northern part of Newton Abbot with the daily freight from Heathfield on 15 July 1971. Headcode 6B15 relates to the 0300 Exeter St Davids to Barnstaple parcels and freight, not this working. *G. F. Gillham*

Kingswear Branch

Right: Judging from the people waiting on the other platform at Torquay an up train must be due, but the down platform can only host No D6334 passing through the shadows at 1025 with an engineers' train in 1960. *Colour Rail (DE2139)*

Right: No D6333 passes Torquay gas works on 2 August 1961 with a Manchester to Kingswear train in tow. Although this would have to be classed as an express, the engine only carries a local train disc. Bearing in mind the old Great Western propensity for upgrading everything possible to express classification, this seems out of character. *Ian G. Holt*

Left: Paignton station sees No D6334 in blue livery with what is shown as 1V57, which is the 0050 Manchester Piccadilly to Penzance parcels, so presumably a portion for the Kingswear branch was dropped off at Newton Abbot. The date is 2 August 1969. Note that the engine still carries a 'D' prefix to the number, even though it has been repainted in corporate blue. *N. A. Hunt*

Left: No D830 *Majestic* is the engine fitted with two Paxman diesels rather than the standard Maybach engines, but this was not pursued with any other 'Warships'. It is seen here climbing away from Paignton with the 'Torbay Express' on Sunday 30 August 1964. 2C91 does not seem an appropriate code for such an important train, which should have read 1C75, unless there had been an engine change at Newton Abbot and the crew hadn't bothered to change the previous code on the locomotive. *G. F. Gillham*

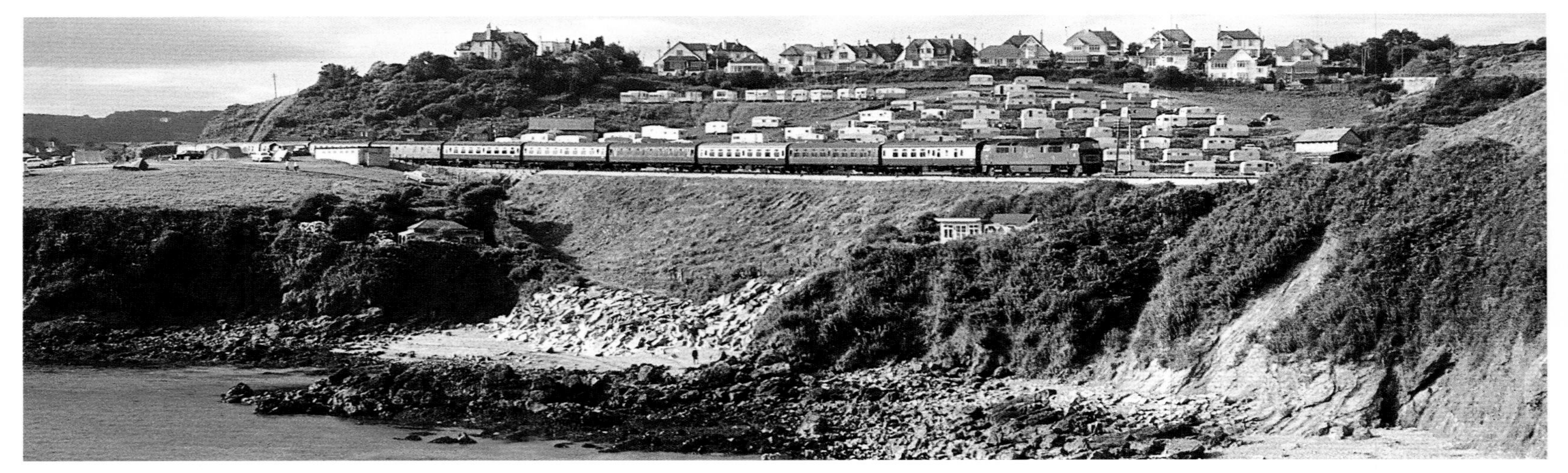

Above: No D1009 *Western Invader* hugs the cliff edge at Waterside between Churston and Goodrington, as it heads 1E40, the 0852 Kingswear to Leeds service, on 23 August 1969. Corporate blue coaches are now becoming more common. The caravans on the hillside give their occupiers a good view of the line! *Bernard Mills*

Left: Most photos at Churston are taken from the road bridge looking towards Kingswear, so this view gives a different perspective. On 7 April 1964, No D6327 is entering the station with the Kingswear portion of the down 'Cornish Riviera' which would have been detached at Newton Abbot. The track of the Brixham branch is still in situ, but all services had ceased in 1963. *Bryan Gibson*

Left: On 4 August 1964, No D7079 has the road leaving Kingswear with 2C90, the 1710 Sunday service to Taunton. Apart from the fact that the line is preserved, and motive power is no longer a 'Hymek', the scene has hardly changed.
R. E. Toopz

Left: Most views at Kingswear are taken of up trains, so it is good to record a Class 22 arriving under the footbridge with a local train from Exeter. Stabling sidings can be seen on both sides of the main line.
Classic Traction

Kingsbridge Branch

Right: Gara Bridge sees No D6333 enter with the 1422 Kingsbridge to Hackney freight on 9 August 1963.
Bryan Gibson

Right: On 3 September 1963, No D6333 is with the 0630 Hackney to Kingsbridge branch freight train at Brent. The station, goods shed, sidings, engine and wagons have all disappeared in accordance with the edicts of the Beeching plan.
Bryan Gibson

Right: No D6329 passes Coombe Farm with the 1422 Kingsbridge to Hackney freight. This picture taken on Wednesday 31 July 1963 illustrates the long-gone pleasures of a rural branch line on a warm summer day.
Bryan Gibson

Right: No D6301 is at Stentiford Hill near Kingsbridge with the 1025 to Paddington on 15 September 1962, replacing what was normally a single-car DMU working. Five coaches and an express headcode make a dramatic change for this branch!
Bryan Gibson

No D6330 leaves Kingsbridge on 23 June 1962 with the 0910 to Paddington which will join up with an express at Brent. From the headcode discs, the train doesn't know whether it is an express or a local! *Bryan Gibson*

It is 16 October 1963, one month after the official closure of the Kingsbridge branch, and No D6310 has been sent down to clear the line of any remaining goods wagons. It is seen here leaving the old station area for the last time. *Bryan Gibson*

Tavistock Junction–Launceston

Right: On 31 December 1965, No D6317 stops at Lydford with the 0800 from Launceston which will have shunted the yard serving the Ambrosia Rice factory at Lifton. The GW branch through Tavistock South had closed at the end of 1962, so this train will take the Southern line throught Tavistock North and Plymouth North Road to reach its destination at Tavistock Junction yard. *Bryan Gibson*

Looe Branch

Right: A shot of a diesel hydraulic on the Looe branch is pretty rare, so this picture of what is probably No D6319 at Coombe Junction is well worth a look. No other details are available. Of course, in recent times, freight traffic to Moorswater is of a different dimension altogether. *Classic Traction*

Coming round the curve into Bodmin Road, as it was then, is No D6342 with the 1057 from Padstow. The date is April 1963, but the view is largely unchanged thanks to the preservation enthusiasts. This shot shows the design of the class to advantage, and I always liked the look of these little engines. *David Cable*

Right: BR blue No D6328 has a weedkilling train in hand at Bodmin General on 16 May 1971, a view which under preservation still looks very much the same. Of interest is that the locomotive still has a 'D' number prefix, although by that date reliveried engines should have dispensed with it. *Bryan Gibson*

Right: No D6314 has arrived at Bodmin General with a train from Bodmin Road in April 1962. The tail lamp is positioned for use after the locomotive has run round its train in order to proceed to Padstow. Note the trolley on the platform and the lorry loading on the left of the picture. *Colour-Rail (DE2318)*

An interesting view from Fowey station of the yard layout, with a pair of 'D6300s' waiting to leave for St Blazey via Pinnock with clay empties. The date is September 1962, but alas everything has long gone, even perhaps the track crews, who at that time certainly never worried about high-visibility clothing. *Colin Judge*

Right: Pioneer 'Warship' No D800 *Sir Brian Robertson* descends the bank from Pinnock tunnel into Fowey station with a load of china clay for unloading at Carne Point on 23 April 1968. The single platform, signalbox and various signals are well displayed, and the car park hosts not only a green Morris 1000 but also an early example of white van man! The line has now been converted into a road, and the whole station area has been redeveloped.
Bernard Mills

Right: An unidentified Class 22 stops to have the brakes pinned down on the clay wagons near Pinnock tunnel on its way from St Blazey to Fowey. A 1C51 headcode is perhaps somewhat optimistic for such a train, but at least the twin headcode panels are being used.
Keith Bachelor

No D600 *Active* pauses in St Blazey yard with a train of vans on an unknown date. The engine looks nice and clean, so possibly the photo was taken shortly after the engine was allocated to St Blazey shed and they regarded it as one of their pets at this time. *Keith Bachelor*

Right: An unidentified 'Warship' has arrived from Fowey via Pinnock with empty clay wagons and has been given the signal to reverse into the yard at St Blazey. The 6C35 headcode has not been specifically identified. *George Hammett*

Right: An against-the-light shot of No D601 *Ark Royal* highlights the amount of fittings on the front end of this original 'Warship' class. By this time, April 1963, the class had been relegated to china clay duties, and in this view the train is seen approaching St Blazey on its way down from the operations around Bugle and Roche. *David Cable*

Above: On 12 August 1967, maroon-liveried 'Warship' No D840 *Resistance* speeds past Luxulyan station with Newquay to Paddington express 1A60 before it descends the bank to St Blazey. The island platform serves only one track now, but the old GWR pagoda shelter stands defiant, as does the disused water tower. The locomotive carries OHE warning labels on its nose, and one blue and grey coach interrupts the neatness of the consist. China clay tips are on the horizon. Note the old water column by the telegraph pole on the right. Various ramshackle buildings complete a view full of interest. *Exe Rail*

Right: The signalman's Vespa scooter leans against the wall of the signalbox at Goonbarrow which is about to be passed by a Class 22 hauling the 1045 Newquay to Par local service of compartment stock on 21 September 1960. The box roof carries a good array of chimneys and ventilators. To the left of the picture, behind the box, can be seen part of the construction of new plant at the china clay facilities which are still in use today. *George Hammett*

Below: Now preserved No D1015 *Western Champion* is seen in a shot taken after the time period for this book but sufficiently interesting to justify inclusion. 1M54, a midday Newquay to Nottingham train, is crossing the notorious bridge over the A30 on Tregoss Moor in 1976. This bridge has been repeatedly hit by incompetent lorry drivers who fail to read the height restriction. However, the new A30 dual carriageway now avoids the bridge, so the DMUs and the few summer weekend HSTs can now pass safely over it. The engine could certainly do with a good clean. *Bernard Mills*

Left: St Dennis Junction is where the line to Burngullow splits from the main branch from Newquay to Par via Luxulyan. A Newquay to Par stopping service passes the junction on 4 October 1961 behind No D6320, with semaphores abounding. *Kidderminster Rail Museum (069162) / P. J. Garland*

Left: On 7 October 1959, an express is preparing to leave Newquay, probably for Paddington. No D6327 is piloting No D826 *Jupiter*. The holiday season is almost over and the crowds will be returning home. *Richard Lewis*

Newquay–Chacewater

Locomotive-hauled trains on the line through Perranporth were extremely elusive for photographers, so this view of No D6320 at Tolcarne Junction, on the approaches to Newquay, with its train from Chacewater is well worth including. The curious signal on the right is intriguing — it was used to control access to the third side of the triangle that was used to turn tender locomotives serving Newquay *Kidderminster Rail Museum (069160) / P. J. Garland*

Falmouth Branch

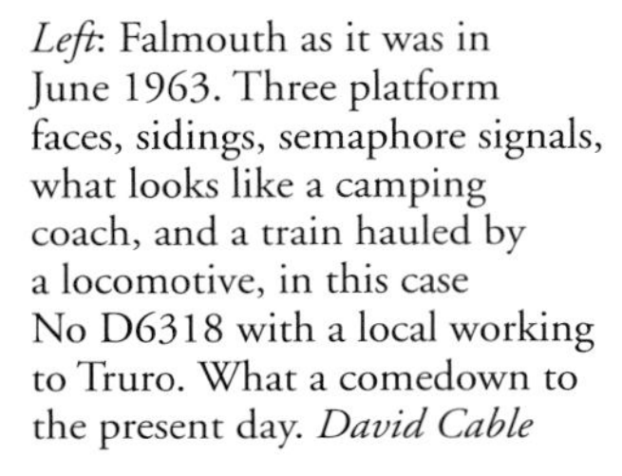

Left: Falmouth as it was in June 1963. Three platform faces, sidings, semaphore signals, what looks like a camping coach, and a train hauled by a locomotive, in this case No D6318 with a local working to Truro. What a comedown to the present day. *David Cable*

Left: No D1001 *Western Pathfinder* is leading a scrap train at Falmouth in 1967. In contrast to the previous photo, the station and sidings have been reduced to a single platform line. *Maurice Dart*

Helston Branch

Right: With a whiff of exhaust smoke, a Class 22 brings the branch train from Helston into Gwinear Road shortly before the line closed in 1966. The crew is about to hand over the token to the signalman, who has cleared the main line for a down train so a comfortable connection should be made. *Eric Irons*

Right: On 24 July 1962, Helston hosts two Class 22s: No D6317 on a local freight and No D6311 with the branch train to Gwinear Road. Both were Laira engines at that time, and to get the two together at such a remote location was indeed fortunate for the photographer. *David Hall*

2C61 is the branch train from Helston to Gwinear Road, and No D6332 waits patiently for departure time. The station still has the old GWR nameboard and shows off the convenience and barrow. The sidings indicate that not much other traffic has been running by this date in August 1962. Note the Class 22, devoid of a yellow front end, and coaches in the carriage siding waiting to take the next service. *Colour-Rail DE2069*

Making a pleasant change from the normal DMU fare, Nos D6319 and D6303 bring the up 'Cornish Riviera' from St Ives along the ledge near Carbis Bay in July 1965, giving passengers glorious views across the Hayle estuary on a lovely summer day. *Colour-Rail (DE2068)*

Southern Main Line to Ilfracombe

Left: No D803 *Albion* has 3C48, the 0542 Seaton Junction to Westbury milk train, in tow at Crewkerne on 8 May 1965, in the days when the main LSWR line to Exeter was still double tracked. The 6-wheel van behind the engine looks like one of interest to the rolling-stock connoisseurs. *Maurice Dart*

Left: This picture, taken on Honiton bank on 29 November 1970, shows an up express with a 1O20 headcode indicating it is heading for a Southern Region destination, although the photographer has stated the train was 1V09, which seems illogical. 'Hymek' 7044 is piloting 'Warship' No 820 *Grenville*, and since 'Hymeks' were far from common on this line, the picture is well worth a look. *Bernard Mills*

Right: Honiton station is host to No D824 *Highflyer* on 11 June 1967 with a Waterloo-bound express. Although both platforms are used, electric signals indicate that bi-directional working can be operated. George Blay's factory peers over the footbridge, while the car park holds a red bus and white car. *Exe Rail (4519)*

Right: Cowley Bridge as seen with the Southern service from Brighton to Plymouth. 'Warship' No D809 *Champion* starts to accelerate on its way from Exeter St Davids in May 1965, with the train still comprised of green Southern Region stock. This train provided restaurant car service as far as Exeter, but those in the know who joined the train at St Davids could sit in the restaurant car and help the staff consume all the leftovers from afternoon tea at the standard price. No wonder my suits were designated as portly in those days! *David Cable*

Right: This picture is undated, but as can be seen, the old Southern main line to North Devon, Plymouth and Cornwall has now been singled at Cowley Bridge Junction, although the signalbox still earns its keep as of old. No D813 *Diadem* has come down from Meldon Quarry with ballast working 7B60. This traffic continues today, but not behind 'Warships'. *John Cooper-Smith*

Below: Approaching Cowley Bridge from Newton St Cyres is this 3C15 train from Torrington to Exeter Riverside, where the wagons will be added to the St Erth to Kensington Olympia service. The locomotive is No D6321 and is seen on 7 May 1967. It always seemed strange to me that the brake van was so often positioned next to the engine on these trains and not at the rear. *Michael Mensing*

Right: On 18 July 1959, the 1445 Plymouth to Exeter Central is seen leaving Crediton behind a pair of Class 22s, namely Nos D6302 and D6303, working in multiple. *S. C. Nash*

Right: One of the first 'Warships', in filthy condition, approaches Crediton hauling the Plymouth to Brighton express in the 1960s with a full rake of Southern stock. *Eric Irons*

Left: No D1020 *Western Hero* brings a ballast train from Meldon Quarry along the singled line from the Okehampton direction on 16 May 1972, to run alongside the line to Barnstaple at Yeoford before merging into up and down tracks at Crediton station. Rationalisation of tracks can be seen from the trackbed remains. The 1V headcode cannot be correct for such a service. *Bernard Mills*

Left: Portsmouth Arms has lost its second track by this day in spring 1968, where this Ilfracombe to Exeter train is passing in the rain. The locomotive is No D812 *The Royal Naval Reserve 1859-1959*, the other 'Warship' apart from No D800 not to be named after a vessel, although it was originally designated to be *Despatch*. *Terry Gough*

No D6333 has been seen several times in this book and has now acquired corporate blue livery whilst still retaining a 'D' prefix to its number. In this view it is hauling the milk train from Torrington to Exeter Riverside near Umberleigh on 19 September 1970. *Exe Rail (877)*

No D6322 is ready to leave Barnstaple Junction for Exeter Central with a five-coach train of Bulleid stock in July 1964. The local train headcode is correct. Is the man in the six-foot tapping wheels *à la* Will Hay as in the film *Oh, Mr Porter!? Colour-Rail (DE2391)*

Left: This view of 'Hymek' No D7084 leaving Barnstaple Junction with a train for Ilfracombe shows the layout with the Torrington branch in the foreground. Whence the train emanated I do not know; either Exeter or Taunton. The date was May 1965, Note that the driver is about to take the token to proceed over the River Taw bridge. *David Cable*

Below: Many shots have been taken of trains on the bridge over the River Taw between Barnstaple Junction and Barnstaple Town, but usually from the Junction end, so this shot restores the balance. On 18 August 1964, No D6316 is taking a train composed of Southern stock to Ilfracombe, so has obviously come up from Exeter and possibly Waterloo. *Kidderminster Rail Museum (018850) / V. R. Webster*

Left: Heddon Mill is on the 1 in 60 climb from Braunton to Mortehoe, and on 1 September 1968 sees No D7068 working hard with its train from Exeter to Ilfracombe. *H. Wells*

Left: One of the original three 'Warships', No 802 *Formidable*, pulls away from Mortehoe & Woolacombe station with local service 2C91 from Ilfracombe to Kingswear. Note the coal by the siding, which judging from the rust is out of use but has what seems to be a rather unnecessary bend in it. Having been at a school which was evacuated to Mortehoe during the war, I can vouch for the remoteness and long walk to civilisation required from this station, especially for a 12-year-old well laden with luggage at the start of the term in January! *Bernard Mills*

Right: 2C87 is the 0735 Newton Abbot to Ilfracombe seen arriving at its destination behind No D820 *Grenville* on 19 August 1968. No doubt the brake blocks are hot after descending the 1 in 36 from Mortehoe. Note the absence of signal arms on the starter signal post. *M. Edwards*

Right: The Ilfracombe carriage sidings have obviously seen better days judging from the rust on them, but at least the station still sees traffic. Corporate blue now rules on 24 August 1968, with 'Hymek' No D7036 waiting to depart. *Bernard Mills*

Barnstaple–Torrington–Halwill Junction

In the first half of the 1960s, I worked for Clarks Shoes. On this day in March 1965, I had to visit their factory in Barnstaple and on the way called in at Torrington, where this misty early morning shot shows No D6307 waiting to depart with a stopping service to Barnstaple Junction. Obviously the steam heating boiler was working that day! Passenger services on this line ceased later that year. *David Cable*

Right: Looking like an 00 gauge model, a Class 22 brings a train of ball clay from Meeth towards Torrington on 23 March 1967. This view was taken from the A386 Hatherleigh to Bideford road, but even with 20/20 vision, the engine number is too small to see. *Bernard Mills*

Right: No D6334 is in charge of this Meeth to Torrington ball clay working, ambling down the hill past Dunsbear platform on an unknown date. Passenger services between Halwill Junction and Torrington had ceased on this line in 1965, although this part of the line stayed open for this traffic until 1982.

Okehampton–Plymouth

Left: In the days when there was an opportunity to divert trains to the West Country away from the sea wall at Dawlish, No D1057 *Western Chieftain* speeds through Lydford on its way to Plymouth. The date is 22 October 1967. The Great Western branch to Launceston can be seen in the background. The footbridge is one of Exmouth Junction's finest, but the platforms are starting to deteriorate. *Exe Rail* (3552)

Left: The Meldon Quarry to Bere Alston section closed in 1968, but track was still in existence in 1969, as we see here on 29 October at Lydford. No D838 *Rapid* is seen with a demolition train ready to leave for Plymouth. The engine name is hardly appropriate for such a duty! *Bryan Gibson*

Right: Maroon-coloured 'Warship' No D817 *Foxhound* gleams in the winter sun as it pulls away from Tavistock North with the daily Plymouth to Brighton express in February 1967. This and the return working were the only trains to carry a restaurant car on this line. Another location now lost to the enthusiast. *Bernard Mills*

Right: No D1072 *Western Glory* enters Tavistock North on 22 October 1967 with a northbound train. The lowering sun creates shadows which highlight the station awnings and footbridge, but no passengers are in sight so the train may terminate here. *Bernard Mills*

Left: The 1146 Plymouth to Waterloo is seen arriving at Tavistock North on a typical cloudy day in December 1963, the 23rd to be precise. The three coaches will certainly be augmented at Exeter if not at Okehampton. Note that the standard Southern headcode of two vertical discs should be used on services such as this. *Bryan Gibson*

Left: Although this shot is rather shadowy, it is included not only because it shows a green 'Western', No D1004 *Western Crusader*, but the train is the notorious passenger, mails and parcels service from Plymouth to Eastleigh, which I believe stopped at every station except Sutton Bingham — what had it done to miss out? Were any passengers brave enough to do the whole trip? The train is on the outskirts of Tavistock in May 1965. *David Cable*

Right: The headcode for this Plymouth to Tavistock North evening local service is wrong as No D861 *Vigilant* ambles towards its destination with a load hardly likely to tax its powers. Note the ER Thompson coach behind the engine, somewhat out of place in a train on such a duty. *David Cable*

Right: I am not going to pretend I was here on official business, because this is the north end of Shillamill tunnel between Bere Alston and Tavistock North. No D6320 emerges with the 1425 service from Plymouth to Exeter Central in April 1963, shortly before steam finished on the through Brighton and Plymouth expresses. *David Cable*

Although the 'Hymeks' were quite frequent operators on the Southern lines to North Devon, they were far from common on the lines through Okehampton. So this shot is very much of interest, showing No D7027 at Bere Alston with the 1405 Plymouth to Waterloo on 6 April 1964. The Callington branch veers off to the right, but no traffic is around at present. *Bryan Gibson*

Right: On an unrecorded misty day, No D6335 approaches Bere Alston with a weedkilling train from Plymouth. Headcode 5P07 does not help, but this train normally only worked once per year. *Bernard Mills*

Right: This is a shot which I have to admit is a cheat. The original, taken on Perutz film, has faded, but in particular the shadows of the Tamar road bridge fell right across the front of the train. Through the miracles of modern science we can now see what I had hoped to portray, namely an unidentified 'D6300' about to pass under the bridge at Ernesettle with a Waterloo to Plymouth express in February 1964. *David Cable*

The footbridge at Devonport Kings Road is not what one might call aesthetic and does little to enhance No D6344 pausing at the station with the 1600 Plymouth to Waterloo. The date is 12 August 1963, and the centre roads look pretty well abandoned. *Bryan Gibson*

Right: Boscarne Junction in June 1962, where No D6319 is shunting a freight which has arrived from Bodmin North. The signal guards the exit from the Wenford branch on to the 'main' line from Bodmin to Wadebridge and Padstow. *Colour-Rail (DE1505)*

Above: Whitsun 1962 at Boscarne Junction sees No D6338 passing with a train to Wadebridge from Bodmin General. The line to Wenford Bridge goes off to the left by the mini signal cabin. Signals are of both Southern and Great Western origin. *Roger Holmes*

Right: No D6341 is at a location known as Guineaport, where the lines from Boscarne Junction and the North Cornwall line run parallel before converging at the approaches to Wadebridge. The 9B04 headcode applies to Bodmin goods trains and is obviously incorrect, but the train may be an unadvertised school service. *Maurice Dart*

A train for Padstow from Bodmin Road is approaching the bridge over Little Petherick Creek. The engine is No D6337 on an unknown date. No doubt the driver couldn't be bothered to climb up to set the correct headcode for a local train, i.e. at the top, but such is life in a backwater!
Colour-Rail (DE2140)

The 0816 from Padstow to Bodmin Road hugs the bank of the River Camel on its way towards Wadebridge behind No D6314 in July 1966. The three-coach train has what seems a superfluity of first class accommodation for such a service. *Colour-Rail (DE2011)*